# BRIDGES
*of the* WORLD

*For master bridge builders, blacksmiths, carpenters, bricklayers and hemp workers. To all those who build bridges and those who cross them.*

# BRIDGES
## *of the* WORLD

GIANCARLO ASCARI
& PIA VALENTINIS

TRANSLATED INTO ENGLISH
BY KATHERINE GREGOR

**OH** EDITIONS

*Absence is neither time nor a road.*
*Absence is a bridge between us,*
*thinner than a hair,*
*sharper than a sword.*

Nâzim Hikmet
*Love Poems*

# CONTENTS

U BEIN BRIDGE
*Amarapura, Myanmar*

THE STONE BASTEI
BRIDGE
*Saxon Switzerland, Germany*

BLACKFRIARS BRIDGE
*London, United Kingdom*

BROOKLYN BRIDGE
*New York, USA*

SNOW BRIDGE
IN PIZ PALÜ
*Sondrio, Italy*

FORTH BRIDGE
*Edinburgh, Scotland*

TOWER BRIDGE
*London, United Kingdom*

THE GLENFINNAN
VIADUCT
*Glenfinnan, Scotland*

ROYAL GORGE BRIDGE
*Colorado, USA*

WEARMOUTH BRIDGE
*Sunderland, United Kingdom*

THE GOLDEN GATE
BRIDGE
*San Francisco, USA*

BRIDGE OVER
THE RIVER KWAI
*Kanchanaburi, Thailand*

THE CELLO BRIDGE

BRIDGE OF BOATS OVER
THE RIVER PO
*Rovigo, Italy, circa 1950*

THE KINTAI BRIDGE
*Iwakuni, Japan*

MOON BRIDGE
*Taipei, China*

A HIGH WIRE BETWEEN
TOWERS
*New York, USA*

THE BRIDGE OF TIME
*Sun City, South Africa*

LIVING ROOT BRIDGE
*Meghalaya, India*

THE BUXIAN BRIDGE
*Huangshan, China*

THE GALATA BRIDGE
*Istanbul, Turkey*

THE PETRONAS TOWERS
SKYBRIDGE
*Kuala Lumpur, Malaysia*

THE AKASHI KAIKYO
BRIDGE
*Kobe, Japan*

PYTHON BRIDGE
*Amsterdam, Netherlands*

THE INFINITE BRIDGE
*Aarhus, Denmark*

NYLON BRIDGE
BETWEEN NEIGHBOURS
*Marseille, France*

Located in the Sahara desert, this arch is one of the largest natural bridges, 122 metres (400 feet) tall and with a 76-metre (250-feet) span. Aloba Arch is one of the most majestic but least visited bridges in the world, because it is so inaccessible.

# NATURAL ARCH

## ALOBA ARCH, CHAD

# PONT *du* GARD

VERS-PONT-DU-GARD, FRANCE

This is the highest Roman aqueduct bridge, built sometime between 20 BC and 60 AD as part of an aqueduct that connected Nîmes to the source of the river Eure, near Uzès. Made up of dozens of arches on three levels, it stretches over the river Gard. It is 49 metres (160 feet) high, 275 metres (902 feet) long and slopes slightly in order to allow the water to flow. After the aqueduct fell into disuse, it became a thoroughfare bridge. During the Middle Ages, you had to pay a toll to the local lords in order to cross it. Napoleon, who was a huge admirer of all things Roman, ordered its restoration in 1850.

Romans were great builders of bridges: just like roads, these became tools for managing and controlling the empire. Using timber, stone, boats and ships, they erected them over the Rhine, the Danube, in the Alps, in Asia and in Africa. The construction of Pons Sublicius, the oldest bridge in Rome, gave birth to the office of *pontifex* (pontiff), the builder of bridges, the highest religious rank to which a Roman citizen could aspire. The title of *pontifex* went from patricians to plebeians, then to emperors until, by the end of the 4th century, it was adopted by the bishops of Rome.

# PONTE SANT'ANGELO

ROME, ITALY

In 134 AD, Emperor Hadrian built a bridge from peperino and travertine stone that led to his mausoleum. Over time it has been frequently restored and renovated by artists and sculptors. In 590 Pope Gregory I had a vision of the Archangel Gabriel, after whom the bridge is named, which inspired the statues of angels made by Bernini and his pupils, which were added in 1669. From the Year of Jubilee 1500, Ponte Sant'Angelo also became a place of executions: it is where the bodies of those who had been put to death were displayed. Starting from 1796, for 68 years, the executioner was Giovanni Battista Bugatti, who carried out his work on the bridge 514 times, but he actually made a living from dyeing umbrellas in a workshop.

# *The* DEVIL'S BRIDGE

## BORGO A MOZZANO, LUCCA, ITALY

Built in about 1100, this bridge is one of
many scattered throughout Europe with
this name. What they have in common is
a bold architectural structure and the
legend of a pact with the devil that
allowed it to be built. According to
legend, a master builder was running late
with his work and the devil promised to
complete it in one night but demanded in
return the first soul that crossed the
bridge. In the end, the master builder
outwitted the devil by getting a dog to
run ahead. The animal's ghost keeps
appearing on moonless nights, chasing
after the man who sacrificed it. The
name given to these constructions is
paradoxical: the word devil comes from
the Greek *diàbolos* (he who divides),
whereas bridges connect.

# The ADMIRAL'S BRIDGE

### PALERMO, ITALY

Built in about 1131 in order to cross the river Oreto, this bridge was commissioned by George of Antioch, a Byzantine admiral in the service of King Roger II of Sicily. He was a highly educated man who knew Greek and Arabic and looked after the economic and political interests of the Kingdom of Sicily in its dealings with other Mediterranean countries.

After repeatedly bursting its banks, the river was diverted in 1938 and nowadays the bridge stands over a garden.

# PONT
# SAINT-BÉNÉZET

AVIGNON, FRANCE

Legend has it that in 1177, urged by an angel, the shepherd Bénézet went to seek permission from the king of France to build a bridge over the Rhône. To show his divine backing, Bénézet lifted a large boulder and threw it into the river. The king was persuaded, and the shepherd completed his work in 1185. A section of the bridge collapsed after the Rhône repeatedly burst its banks, but the boulder is still there and remains a point of interest on local tours.

At 205 metres (669 feet) long, this is considered Europe's oldest covered bridge. It crosses the river Reuss and connects the two oldest parts of the city of Lucerne.

Its name comes from its proximity to St Peter's Chapel, where you can see 17th-century wooden panels depicting the history and principal events of the city. The *Wasserturm* (water tower) stands in the middle of the bridge. It is an octagonal tower erected for defence purposes and is one of the city's landmarks.

# KAPELLBRÜCKE

LUCERNE, SWITZERLAND

# A RAINBOW *in a* CAMPIDANO FIELD

SARDINIA, ITALY

*Somewhere over the rainbow*
*Skies are blue*
*And the dreams that you dare to dream*
*Really do come true.*

H. Arlen and E.Y. Harburg,
*Over the Rainbow*

Standing with our backs to the sun in the field, we can see a rainbow. If the sun's rays go through ice crystals in a particular direction, the rainbow may appear upside down: its colours are very deep and it is a rare event that typically occurs in polar regions.

# PONTE
# DELLE TORRI

SPOLETO, ITALY

*I travelled up to Spoleto and went on the aqueduct that also acts as a bridge between two hills. Its 10 brick arches that dominate the entire valley have been standing there calmly for centuries, while the water still gushes from one end to the other.*

J.W. Goethe, *Italian Journey*, 1786–1788

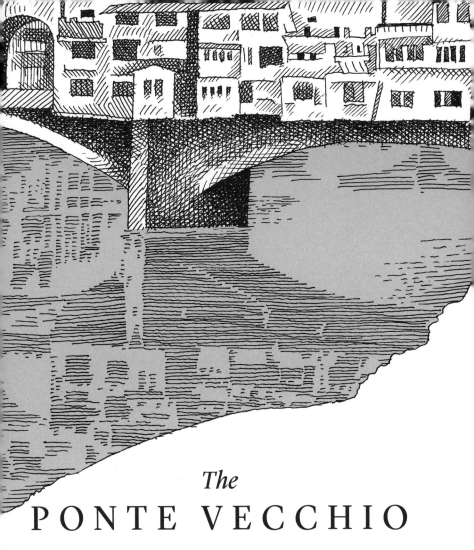

# *The*
# PONTE VECCHIO

### FLORENCE, ITALY

Built in its current form in 1345, its design is attributed to Taddeo Gaddi and it was the West's first bridge with segmental arches. It has withstood many floods and survived the German bombings in 1944. The bridge acts as a continuation of Via Guicciardini, with shops on both sides.

At first, there were butchers, poultry shops and greengrocers, who threw their waste into the Arno. Eventually, the Vasari Corridor was built above the bridge and, around 1593-95, repelled by the smell of meat, Ferdinand I de' Medici decided to replace the food shops with goldsmiths and silversmiths, which are still there to this day.

When building work on the bridge began in 1357, egg whites were mixed with mortar in order to make up a material that could withstand the river Vltava when it overflowed. Charles IV, who commissioned it, apparently got the residents of villages in Bohemia to send large quantities of eggs. The inhabitants of one village, Velvary, sent their eggs hard boiled, as they wanted to make sure they didn't break. The construction lasted 45 years, and in 1378 the king's funeral procession crossed the almost-completed bridge.

Legend has it that at night the statues of the saints – called "stone protectors" – that adorn the bridge come to life to protect the newborn babies on the nearby island of Kampa.

# CHARLES BRIDGE

PRAGUE, CZECH REPUBLIC

# MOSTAR BRIDGE

MOSTAR, BOSNIA AND HERZEGOVINA

The city of Mostar is split in two by the river Neretva and is named after the bridge which connects the two sides of the river: the Stari Most (Old Bridge).

Built in 1557 by order of Sultan Suleiman the Magnificent, it was destroyed in 1993 by Croat-Bosniak forces during the Bosnian war. The bridge became a symbol of an encounter between worlds, cultures and religions, ruthlessly divided by that war.

Rebuilt in 2004, it is a UNESCO World Heritage Site. An annual diving competition has been held since 1968, which was seen as a rite of passage for 16-year-old boys: the dangerous 24-metre (78-feet) dive into the Neretva from Mostar Bridge. These days, it's open to professional divers and thrill seekers alike.

# *The* RIALTO BRIDGE

VENICE, ITALY

Venice's oldest landmark is the Rialto bridge, where the city's market was transferred in 1097. The stone bridge, which replaced its timber predecessor, is one of the four over the Grand Canal and was completed in 1591 by the architect Antonio da Ponte. He is also rumoured to have made a pact with the devil in order to complete the building work.

The bridge, made of a single, 28-metre (92-feet) span, is home to six shops on each side, and because of its proximity to the Zecca – the building where the Republic of Venice coins were once minted – it used to be called Ponte della Moneta (the Coin Bridge).

# *The* BRIDGE
# *of* SIGHS

VENICE, ITALY

*I stood in Venice, on the Bridge of Sighs,*
*A palace and a prison on each hand:*
*I saw from out the wave her structures rise*
*As from the stroke of the enchanter's wand:*

Lord Byron

In 1600 the Doge Marino Grimani ordered a bridge to be built in order to create a link between the New Prison and the interrogation rooms. The bridge is enclosed by walls of white stone with windows, behind which prisoners were taken to their cells, sighing. One of the few prisoners who managed to escape was Giacomo Casanova, who wrote of his rocambolesque flight in *Histoire de ma fuite des prisons de la republique de Venice qu'on appelle les Plombs.*

# SHAHARAH BRIDGE

SHAHARAH, YEMEN

Called the Bridge of Sighs, like its Venetian counterpart, this bridge crosses a 300-metre (984-feet) gorge and was built in the 17th century in such a way that it could be quickly demolished in case of a Turkish attack. Prior to its construction, the residents of the two mountain chains had only one way to communicate and trade goods: by undertaking a perilous journey on foot down the mountain, then up.

The village of Shaharah, on the edge of the mountain, lived on produce from the cattle, rainwater and produce from the terraced fields.

The Pont Neuf crosses the Seine and, despite its name, is the oldest bridge in Paris: its construction was completed in 1607. Decorated with 385 grotesque *mascarons* carved in the stone by sculptor Germain Pilon, it was one of the first bridges with pavements and became a place of entertainment for the whole of Paris thanks to performances by musicians, actors, jugglers and acrobats. Frequented by prostitutes, pickpockets and charlatans, it also held gallows and was the scene of robberies and murders. The district where it was erected was, in 1314, the place of execution by burning at the stake of Jacques de Molay, the last Grand Master of the Order of the Knight Templars, by order of King Philip the Fair. In 1985 the artists Christo and Jeanne-Claude wrapped Pont Neuf entirely in paper for 14 days.

# *The* PONT NEUF

PARIS, FRANCE

# TREPPONTI

COMACCHIO, FERRARA

Comacchio is a city in the lagoon of the
Po Delta, which can also be viewed by
boat via the canals. The Trepponti dates
back to 1638 and is built from Istrian
stone according to a plan by Luca Danese
and commissioned by Cardinal Giovan
Battista Pallotta, who cleaned up the
lagoon. The bridge is the intersection of
four canals and allows access into the
city: because of this strategic position,
two turrets for military use were
subsequently added.

# TREE-TRUNK BRIDGE

## SALZBURG ALPS, AUSTRIA

*Perhaps only the trees*
*fully know the mystery of water.*
Jules Renard

# KHAJU BRIDGE

## ISFAHAN, IRAN

Built by the Persian Shah Abbas II in around 1650, Khaju Bridge also serves as a dam on the river Zayanderud and as a building for meetings. The bridge is three storeys high, has 24 spans and is around 130 metres (426 feet) long. The Shah admired the view from the pavilion at the centre of the bridge, and nowadays you can sit and watch the river while sipping tea or coffee in establishments located on the floor below.

A bridge over one of the Ganges
tributaries. In the 18th century it was about
73 metres (240 feet) long and could be put
up and dismantled quickly to allow the
local rajah and people to leave the city in
case of invasion. It rests on timber poles
and is made of ropes and wooden and
bamboo poles. A fakir used to live in the
building at the top of the mountain.

# ROPE BRIDGE

SRINAGAR, INDIA

# PONTE MEDICI

Located near the Brera Academy, from
the 18th century it connected the two
shores of the small lake of San Marco,
where there is currently a square and an
underground car park. From the 14th
century, a network of navigable canals
(the Navigli) crisscrossed the city of Milan
until, in the 1920s, a campaign was
launched for them to be closed. A
newspaper of the time wrote:
"The Naviglio is a social danger because
of the attraction it holds for the weak and
defeated of a large metropolis, and
suicides. It is a public danger on foggy
winter nights for men and old people who
can fall in. Besides, in the new Italian life
stipulated by Fascism, the reasons for
affirming and improving race must have
the upper hand over any other
consideration . . ."

And so most of the Navigli were
covered up and only a few photographs
remain of the perfect simplicity of this
bridge.

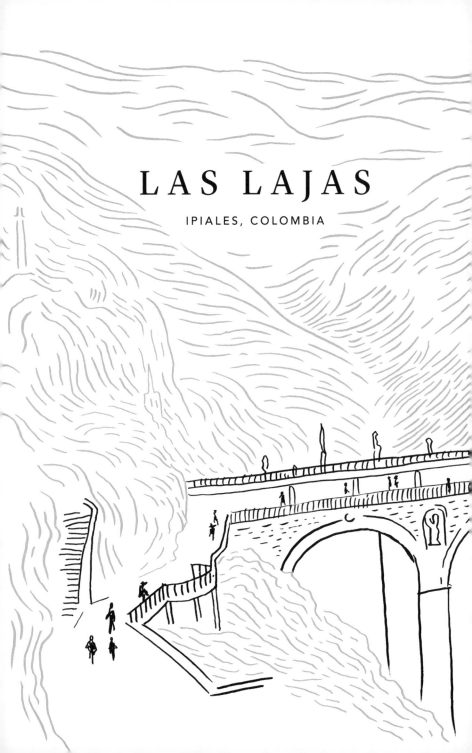

# LAS LAJAS

IPIALES, COLOMBIA

We're in the Andes, 2,900 metres (9,514 feet) above sea level. A hundred metres (328 feet) over the river Guáitara stands Las Lajas, a Gothic cathedral accessible by a bridge with white and grey stone arches. To the hundreds of thousands of pilgrims who come every year, it's a miracle of God suspended over the abyss. Legend has it this is where the Virgin Mary appeared to Rosa, a deaf-mute indigenous girl who was able to speak after the sighting.

# PUENTE NUEVO

## RONDA, SPAIN

After the previous bridge collapsed six years from its inauguration, killing 50 people, Puente Nuevo was completed in 1793. It's as tall as a 30-storey building, made with stone quarried from the bottom of the El Tajo gorge. Above the central arches there is a small room where political prisoners were kept during the Spanish Civil War. The city of Ronda is home to the ashes of Orson Welles, who said, "A man does not belong to the place he was born in, but to the place he chooses to die."

# DRAWBRIDGE
### *of the* FORT *of* PONTA DA BANDEIRA

LAGOS, PORTUGAL

The use of movable bridges began to spread in the 14th century. They were used in various parts of fortresses: the paths taken by the guards on their rounds, the galleries and the entrance to the towers. They were used in ships, on roads, crossings and wherever one needed to quickly prevent or restore passage. Bridges that could be activated by just one person – such as closing openings through which the enemy could be struck – were therefore invented. And when the residents of a castle barricaded themselves in and raised the drawbridge, the besiegers could come in from the top using bridges lowered from mobile towers.

Amarapura was the imperial capital of
Burma – now Myanmar – from 1783 until
King Mindon Min decided to build a new
one in Mandalay, a little further north.
The materials from the Amarapura palaces
and dismantled buildings were reused and
transported by elephant to the new
location. The leftover timber from the
palaces and temples was used for building
the U Bein footbridge in around 1850,
which crosses the Taungthaman Lake.
It is the longest teakwood bridge in the
world and is 1.2 kilometres (0.75 miles)
long. In addition to tourists, the monks
from the Mahagandayon monastery in
Amarapura travel over it on foot or
by bicycle.

# U BEIN BRIDGE

## AMARAPURA, MYANMAR

# *The* STONE BASTEI BRIDGE

SAXON SWITZERLAND, GERMANY

This bridge was built in 1851 over a 40-metre-deep (131-feet-deep) ravine to replace a previous one made of wood. Also visible from Dresden, it's the most panoramic spot in the natural park of Saxon Switzerland. As far back as the 19th century, the area was frequented by many tourists, as well as Romantic painters, poets and composers, such as Carl Maria von Weber, William Turner and Antonín Dvořák. Caspar David Friedrich did many sketches here and in 1823 painted *Die Felsenschlucht* (Rocky Ravine), which presents a partial view of the ravine without the bridge.

# BLACKFRIARS BRIDGE

## LONDON, UNITED KINGDOM

A bridge for road traffic and pedestrians
over the River Thames, Blackfriars was
completed in 1869 and is composed of five
iron arches. Its name derives from a nearby

monastery of Dominican friars (Blackfriars) who wear a black cappa. The bridge acquired a sinister fame in 1982, when Italian financier Roberto Calvi was found hanging from one of its arches: a mysterious death mentioned in the context of mafia interests, secret societies and political intrigue.

# BROOKLYN BRIDGE

## NEW YORK, USA

Finished in 1883, Brooklyn Bridge is over 1.8 kilometres (1.1 miles) long and is the first ever bridge entirely made of steel. For a long time, it was the longest suspension bridge in the world. However, it didn't bring luck to its architect, the engineer Roebling, who was the victim of an accident during construction, or to his son, who brought the work to completion having been left partially paralysed after an injury during the works. Construction was bedevilled by all kinds of misfortunes: strange illnesses, the death of workmen and technicians, and people injured and killed during an overcrowded celebration on the bridge when rumours that it was about to collapse created panic. To reassure New Yorkers that the edifice was safe, in May 1884, the most famous circus impresario of the time, P.T. Barnum, crossed the Brooklyn Bridge with a parade of 21 elephants, 7 camels and 10 dromedaries. A stroke of genius, given how much this bridge has since been loved by American writers and film directors.

# SNOW BRIDGE
## *in* PIZ PALÜ

SONDRIO, ITALY

On glaciers, you can sometimes walk on a
strip of frozen snow that connects one side
of a crevasse to the other.

Opened in 1890, this 2,467-metre-long
(8,094-feet-long) railway bridge was the
first in Britain to have been built entirely
of steel – 53,000 tons of it.

It is considered a wonder of the
industrial age and became a UNESCO
World Heritage Site in 2015.

# FORTH
# BRIDGE

## EDINBURGH, SCOTLAND

# TOWER BRIDGE

LONDON, UNITED KINGDOM

Together with Big Ben, the bridge is one of London's most famous symbols, although it was built relatively recently. Erected in 1894 from stone, granite, cement and steel, it is 244 metres long (800 feet long). It has two suspension footbridges, as well as one for motor vehicles, which is regularly raised in order to allow the passage of boats on the Thames. The bridge was a chocolate colour to start with and was subsequently painted red, white and blue in 1977 to mark the Queen's Silver Jubilee. In 1912 the aviator Frank McLean negotiated his biplane between the lower and upper spans. In 1999 Jef Smith, Freeman of the City of London, crossed Tower Bridge with his sheep, exercising an ancient right.

# *The*
# GLENFINNAN VIADUCT

GLENFINNAN, SCOTLAND

Built in the early 20th century, this viaduct is a section of the West Highland Line that crosses mountains, moors, precipices and cliffs by the North Sea. Traffic on it has substantially increased since it appeared in the *Harry Potter* films – so much so that British Transport Police issued an official notice, warning people not to walk along the rails to take pictures of the film location of the Hogwarts Express. The warning points out that we are not wizards but ordinary muggles.

# ROYAL GORGE BRIDGE

## COLORADO, USA

The bridge was the brainchild of Lon
P. Piper, a Texan who in 1929 came up
with a lucrative way for tourists to enjoy
a spectacular view of the Arkansas River
in Colorado. There is an amusement
park at both ends of the 384-metre-long
(1259-feet-long) bridge, which is the
highest in the United States. It can be
crossed on foot – and by car only when
the parks are closed. The Royal Gorge has
no rigid vertical trellis, which is why it
sways and vibrates when vehicles and
people travel across it. For $30 you can
also bungee jump into the gorge from a
tower over the canyon.

# WEARMOUTH BRIDGE

## SUNDERLAND, UNITED KINGDOM

The first Wearmouth bridge dates back to 1796; this is the third, built in 1929 and 114 metres (374 feet) long. It is arc-shaped and allows the passage of vehicles, bicycles and pedestrians. It is next to Wearmouth Rail Bridge, intended for trains.

# *The* GOLDEN GATE BRIDGE

## SAN FRANCISCO, USA

The entrance into San Francisco Bay from the Pacific Ocean was called *Chrysopylae* (golden door in Greek), or Golden Gate, by John C. Frémont, an engineer who found a resemblance between the Californian strait and the Golden Horn, the estuary that splits the city of Istanbul in two – hence the name of the suspension bridge inaugurated on 27 May 1937 that crosses the bay. The horizontal structure that allows the crossing is suspended from cables held by towers at the ends of the bridge, which has a slightly curved body. Its creation was cosmopolitan: it was planned by Joseph Strauss (the U.S.-born son of German immigrants), designed by Irving Morrow (an American who received his professional training in Paris) and financed by Amadeo Pietro Giannini (the U.S.-born son of Italian immigrants, founder of the Bank of Italy, co-founder of the Bank of America and, some say, the inventor of modern banking). Morrow chose to have the bridge painted orange so that it could be seen even through the fog. The Golden Gate bridge has one of the world's largest number of suicides.

# BRIDGE OVER *the* RIVER KWAI

KANCHANABURI, THAILAND

This bridge is the true protagonist of a novel by Pierre Boulle, and inspired the film by David Lean, which tells its eventful story. It allows a 415-kilometre (258-mile) railway track through the jungle between Myanmar and Thailand and was built in 1943 by 60,000 Australian, British, Dutch and American prisoners of war, as well as 200,000 Asians forced into hard labour. Over 100,000 workers died because of the inhumane conditions so that this enterprise might be completed in only 16 months. Of the original bridge, destroyed by Allied bombs in 1945, only the external wooden spans remain – the very ones the prisoners tried to sabotage by infesting them with termites. We also still have the tune whistled by the prisoners in the film *The Bridge on the River Kwai* to the theme of "Colonel Bogey".

# *The* CELLO BRIDGE

*The 59th Street Bridge Song* – Simon & Garfunkel
*Misty Morning, Albert Bridge* – The Pogues
*London Bridge is Falling Down* – nursery rhyme
*The Bridge* – Elton John
*Erie Canal* – Bruce Springsteen
*The Brooklyn Bridge* – Frank Sinatra
*Get Up (I Feel Like Being a) Sex Machine* – James Brown
*Sur le Pont d'Avignon* – famous French song dating back to the 15th century
*Across the Bridge* – Herb Alpert
*Chelsea Bridge* – Ella Fitzgerald
*Bridge of Pain* – Public Enemy
*Geordie* – folk song
*Seven Bridges Road* – The Eagles
*Sous les Ponts de Paris* – Juliette Gréco

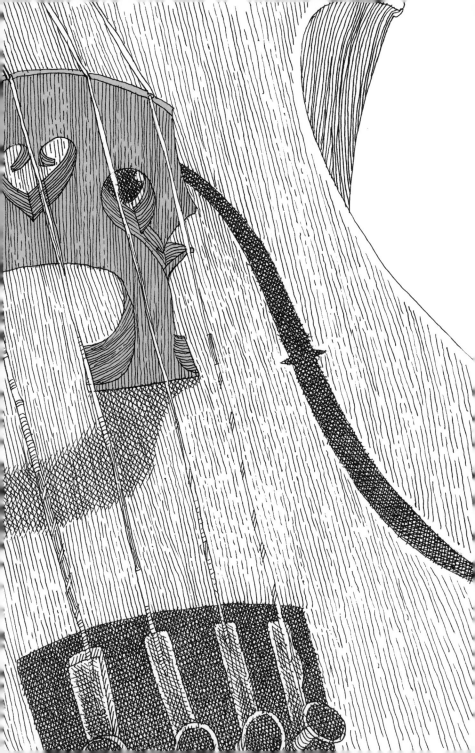

# BRIDGE *of* BOATS OVER *the* RIVER PO

ROVIGO, ITALY, CIRCA 1950

When the water rises, so does the bridge. Easy to disassemble and remove, bridges made with boats are usually temporary and are still used in the present day. They are made during wars to enable troops to cross, and taken apart afterwards to prevent the enemy from using them. Pliny the Elder tells of how ancient Romans built a bridge of boats across the Strait of Messina.

# *The* KINTAI BRIDGE

IWAKUNI, JAPAN

This pedestrian bridge dates back to 1673, and crosses the Nishiki River. Nowadays, it leads to the oldest part of the city and allegedly used to be the entrance into the quarters of the samurai, who were the only ones allowed to cross it. It was originally made entirely of wood, but in 1953, after a powerful typhoon, the bridge was rebuilt with stone pillars to support the five wooden arches. This way, in case of natural disasters, the structure can remain intact.

# MOON BRIDGE

### TAIPEI, CHINA

Standing over a 13-hectare lake in Dahu
Park, it is designed so that moonlight may
produce a perfect mirror image of the
bridge on the water. The location is like
something from a fairy tale, but it is
actually near Taiwan's industrial area,
amid factories and shopping centres.

# A HIGH WIRE
## *between* TOWERS

NEW YORK, USA

*Whenever carpenters start building a bridge,
whenever magicians produce a string on a stage,
whenever children play at a tug of war
and whenever illegal tightrope walkers set up a high wire,
there's always a moment
when the thread dangles freely between two points, and smiles.*
Philippe Petit

On 17 August, 1974, the Twin Towers in New York were still being built. At 7 a.m., passers-by, mouths agape, looked up at a distant figure walking on a high wire between the two buildings. It was Philippe Petit, a 25-year-old French tightrope walker and juggler, undertaking his extraordinary feat. He had already played similar tricks in Notre-Dame, in Paris and at Niagara Falls, but, as a skilled former street pickpocket on monocycle, he prepared the New York stunt as though it were the crime of the century.

After spending the night hiding in one of the buildings with his assistants, in the morning he shot a steel wire between the two towers using a bow and arrow. He then walked up and down it eight times, cheered by the crowd below, holding a three-metre pole to keep his balance.

At the end of his walk, he was arrested and the district attorney sentenced him to give a performance for children in Central Park. Sadly, his exploit at the Twin Towers can never again be repeated.

# *The* BRIDGE
## *of* TIME

### SUN CITY, SOUTH AFRICA

The Bridge of Time is located at the
entrance to an amusement park known as
the African Las Vegas, which has cinemas,
hotels, restaurants, game areas and
excursions amid rhinoceros and crocodiles.
The Bridge of Time is guarded by stone
elephants and shakes every hour
during a show that simulates a powerful
earthquake, accompanied by the eruption
of a fake volcano.

# LIVING ROOT BRIDGE

MEGHALAYA, INDIA

In north-east India, one of the world's most humid regions, a bridge made of timber would rot in no time at all. So the Khasi people build bridges using the long roots of rubber trees, stretching and shaping them for 15 or 20 years until they reach the river bank opposite. In time, these bridges become increasingly strong and can apparently bear the weight of 50 men. When, during the monsoon season, water prevents travel, these bridges are the only way to cross the forest on foot.

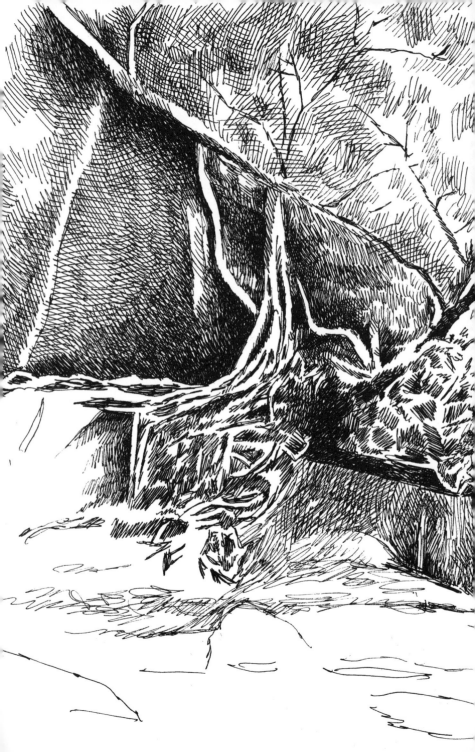

# *The* BUXIAN BRIDGE

## HUANGSHAN, CHINA

This bridge was deliberately designed to look much older than it is, in a spot often visited by impenetrable cloud banks. Although the decorations on the handrails and its shape make it look like an ancient Chinese relic, it was built out of cement in 1987 and crosses the Xihai Grand Canyon (Xihai meaning "West Sea" because of the surrounding ocean of fog). Located in a place known as a "Magic Scenic Area", it quickly earnt the nickname "Fairy Walking Bridge".

# *The* GALATA BRIDGE

## ISTANBUL, TURKEY

The present-day bridge, completed in 1994, is a lift bridge across the Golden Horn estuary and connects the old part of the city and the modern one. The first Galata Bridge was built in 1845, but there had been others in that place. Emperor Justinian I had one erected in the 6th century, and various boat bridges have linked the banks of the Golden Horn over the centuries. In around 1500 Sultan Bayezid II asked Leonardo da Vinci and Michelangelo to design a bridge, but neither of their plans were accepted. A novel by Mathias Énard, *Tell Them of Battles, Kings, and Elephants*, imagines Michelangelo in Istanbul to build the bridge.

*Leonardo da Vinci's single-span design for the Galata Bridge.*

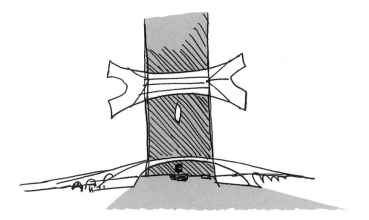

# *The* PETRONAS TOWERS SKYBRIDGE

KUALA LUMPUR, MALAYSIA

This 58-metre-long (190-feet-long), 170-metre-high (557-feet-high) bridge connects the twin towers, named after an oil company and was inaugurated in 1996. They house offices, a theatre and the Petroleum Resource Centre.

Until 2004, the 452-metre (1,482-feet) towers were considered the tallest in the world – but only if the pinnacles are included.

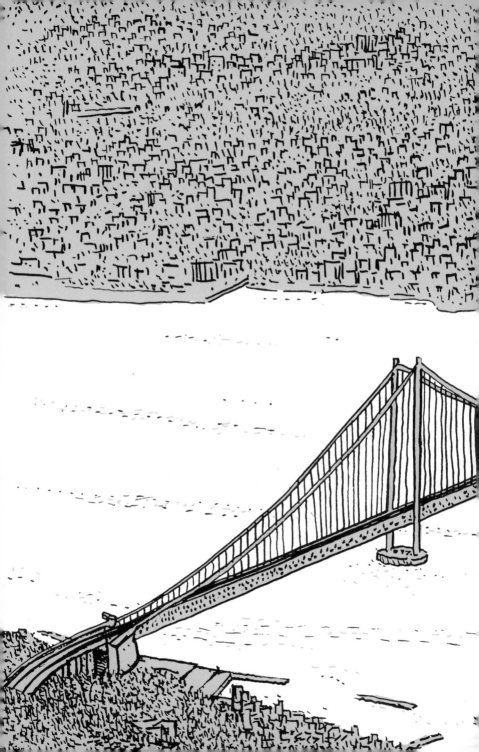

# The AKASHI KAIKYO BRIDGE

## KOBE, JAPAN

The Akashi Kaikyo Bridge is the world's longest suspension bridge – 3,911 metres (12,831 feet) long and 282 metres (925 feet) high – and rests on two concrete cylinders with a 70-metre (229-feet) diameter each. It took 10 years to build and was completed in 1998. Given its dimensions, only 20 per cent of its bearing capacity is intended to carry the traffic on the bridge; the rest needs to support its own weight. It is lit up at night and the colour of the lights changes according to the date and season.

The Python Bridge takes its name from its shape (it was initially also known as the Salamander, the Anaconda and the Dinosaur), but its formal name is Hoge Brug. It crosses the canal between Sporenburg and Borneo Island. Built in 2002 for pedestrians and cyclists, it is 93 metres (305 feet) long and was designed by Adriaan Geuze, who chose its bright red colour.

# PYTHON BRIDGE

## AMSTERDAM, NETHERLANDS

# *The* INFINITE BRIDGE

## AARHUS, DENMARK

As much a sculpture as a bridge, this circular bridge, with a 60-metre (196-feet) diameter, stands astride the beach and the sea. It was conceived and created by the Danish architecture studio Gjøde & Povlsgaard Arkitekter on the occasion of the international Sculpture by the Sea event in 2015. It stands on the site of a pier that no longer exists: a popular spot where people would go to eat, dance and sunbathe.

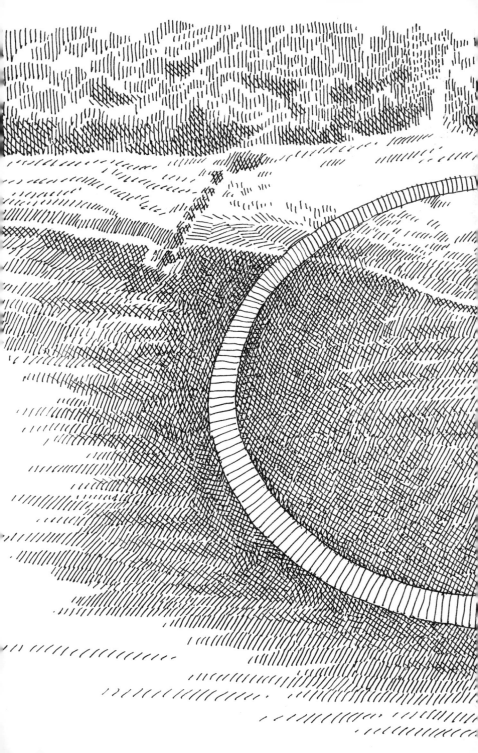

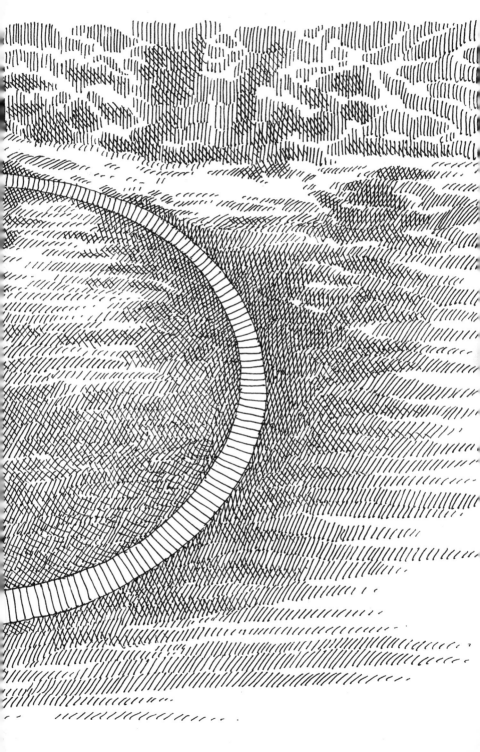

# NYLON BRIDGE
## *between*
# NEIGHBOURS

MARSEILLE, FRANCE

*First, you do the washing.*
Until the last century, people would do their
washing with ash lye and hot water. The laundry
would be soaked overnight, then rinsed in the
fountains.
*Then you hang up the washing.*
Without pegs: the clothes peg is an 18th-century
invention. The kind we use now, with the steel
spring that allows it to be opened and closed, was
invented in Vermont in 1853 by David M. Smith.
*Then there's hoping the weather doesn't change.*
In the Middle Ages, people used to sing:
"No sooner do I start on the laundry than it rains.
If the weather's fair, it suddenly turns dark
With lightning, thunder and air full of spark."

*At 20, all the bad still stretched before us,*
*And all the good we knew we had within,*
*Beyond the bridge lies life, just waiting for us.*
*Beyond the fire lies love, ready to begin.*

*Oltre il Ponte,* words by Italo Calvino,
music by Sergio Liberovici.

Giancarlo Ascari is an architect and comic book author who signs his work with the name Elfo. He has written and drawn for *Linus, l'Unità, Corriere della Sera* and *Diario*. He is the author of the graphic novel *Love Stores* (2005), *Tutta Colpa del '68* (2008), *Sarà una Bella Società* (2012) and *L'arte del Complotto* (2015).

Pia Valentinis was born in Udine and lives in Cagliari, Italy. She writes and illustrates books for children. In 2015 her first graphic novel, *Ferriera*, won the Andersen Prize for best comic book.

This English language edition published in 2022 by OH Editions
Part of Welbeck Publishing Group.
Based in London and Sydney.
www.welbeckpublishing.com

First published in 2018 by Giunti Editore S.p.A/Bompiani, Firenze-Milano
Original title: *Ponti Non Muri*
www.giunti.it
www.bompani.it

Design © 2022 OH Editions
Text and Illustrations © Giancarlo Ascari and Pia Valentinis 2022
English translation © Welbeck Publishing Group

A CIP catalogue record for this book is available from the British Library.

ISBN 978-1-91431-756-9

Publisher: Kate Pollard
Translator: Katherine Gregor
Editor: Matt Tomlinson
Designer for UK edition: Studio Noel
Illustrators: Giancarlo Ascari and Pia Valentinis
Production controller: Arlene Lestrade

Printed and bound by RR Donnelly in China

10 9 8 7 6 5 4 3 2 1